Bacteriology Uncovered

The Secrets of Microbial World Revealed

Bhupen Thapa

Copyright © 2024 Bhupen Thapa
All rights reserved.
No part of this book may be reproduced or transmitted in any form or by any means, electronic or mechanical, including photocopying, recording, or by any information storage and retrieval system, without permission in writing from the copyright holder.

Table of Content

Bacteriology Uncovered 1
The Secrets of Microbial World Revealed 1
   Chapter 1: Introduction to Bacteriology 2
      Overview of Bacteriology 2
      History of Bacteriology 3
      Importance of Studying Bacteriology 4
   Chapter 2: The Microbial World 5
      Classification of Microorganisms 6
      Characteristics of Bacteria 7
      Differences Between Bacteria and Other Microorganisms 8
   Chapter 3: Bacterial Structure and Function 10
      Cell Structure of Bacteria 10
      Functions of Bacterial Components 11
      Cell Division and Reproduction in Bacteria 12
   Chapter 4: Bacterial Growth and Metabolism 14
      Factors Affecting Bacterial Growth 14
      Nutritional Requirements of Bacteria 15
      Metabolic Pathways in Bacteria 17
   Chapter 5: Bacterial Genetics 19
      DNA Replication in Bacteria 19
      Gene Expression and Regulation in Bacteria 20
      Horizontal Gene Transfer in Bacteria 21
   Chapter 6: Bacterial Pathogenesis 23
      Mechanisms of Bacterial Pathogenicity 23
      Host-Pathogen Interactions 24
      Immune Response to Bacterial Infections 25
   Chapter 7: Antibiotics and Bacterial Resistance 27
      Overview of Antibiotics 27
      Mechanisms of Antibiotic Action 28
      Bacterial Resistance to Antibiotics 29
   Chapter 8: Industrial and Environmental Applications of Bacteriology 31
      Role of Bacteria in Industry 31
      Bioremediation and Environmental Cleanup 32
      Probiotics and Other Beneficial Bacteria 34
   Chapter 9: Future Trends in Bacteriology 36
      Emerging Infectious Diseases 36
      Advances in Bacterial Research 37
      Ethical Considerations in Bacteriology 38
   Chapter 10: Conclusion 40

    Recap of Key Concepts 40
    Implications of Bacteriology in Healthcare 41
    Future Directions in Bacteriology Research 42
Dear Readers, ... 43

A Primer for Healthcare Professionals

Chapter 1

Introduction to Bacteriology

Overview of Bacteriology

Bacteriology is the branch of microbiology that focuses on the study of bacteria, which are single-celled microorganisms that can be found virtually everywhere in our environment. These tiny organisms play a crucial role in various biological processes, and understanding their behavior and characteristics is essential for numerous fields, including medicine, agriculture, and environmental science.

In the field of bacteriology, researchers study the structure, function, and behavior of bacteria in order to better understand their impact on living organisms and the environment. By studying bacteria at the molecular level, scientists can gain insights into their mechanisms of action, their ability to cause disease, and their potential for beneficial applications, such as bioremediation and biotechnology.

One of the key areas of focus in bacteriology is the study of bacterial pathogenesis, which involves understanding how bacteria interact with their hosts to cause disease. By identifying the mechanisms by which bacteria invade host cells, evade the immune system, and cause tissue damage, researchers can develop new strategies for preventing and treating bacterial infections.

Another important aspect of bacteriology is the study of antibiotic resistance, which has become a major public health concern in recent years. Bacteria have the ability to adapt and evolve in response to the selective pressure of antibiotics, leading to the emergence of drug-resistant strains that are difficult to treat. By studying the mechanisms of antibiotic resistance, scientists can develop new drugs and treatment strategies to combat these deadly pathogens.

Overall, bacteriology is a dynamic and rapidly evolving field that plays a vital role in our understanding of the microbial world. By shedding light on the behavior and characteristics of bacteria, researchers can develop new tools and technologies to improve human health, protect the environment, and advance our knowledge of the natural world.

History of Bacteriology

Bacteriology, the study of bacteria, has a rich and fascinating history that dates back centuries. The field of bacteriology began to emerge in the late 17th century, with the invention of the microscope by Antonie van Leeuwenhoek. Leeuwenhoek was the first to observe and document the existence of bacteria, which he described as "animalcules" swimming in a drop of water. This groundbreaking discovery laid the foundation for the study of microbes and their role in disease.

In the 19th century, the field of bacteriology experienced significant advancements, thanks to the work of scientists such as Louis Pasteur and Robert Koch. Pasteur's experiments with fermentation and vaccination revolutionized our understanding of bacteria and their impact on human health. Koch, on the other hand, developed a set of postulates to prove the causal relationship between specific bacteria and infectious diseases, a method still used in microbiology today.

One of the most significant discoveries in the history of bacteriology was the identification of penicillin, the first antibiotic, by Alexander Fleming in 1928. This breakthrough revolutionized the treatment of bacterial infections and saved countless lives. The discovery of penicillin paved the way for the development of other antibiotics and antimicrobial agents, leading to further advancements in the field of bacteriology.

In the modern era, bacteriology continues to play a crucial role in healthcare and public health. The study of bacteria has led to the development of vaccines, diagnostic tests, and treatments for a wide range of infectious diseases. Bacteriologists are constantly researching new ways to combat antibiotic resistance and emerging infectious diseases, ensuring that we stay one step ahead of these microbial threats.

Overall, the history of bacteriology is a testament to the ingenuity and dedication of scientists who have worked tirelessly to uncover the secrets of the microbial world. Through their efforts, we have gained a deeper understanding of the role bacteria play in our lives and have made significant strides in preventing and treating infectious diseases. As we continue to explore the world of bacteria, we can look forward to even more groundbreaking discoveries that will shape the future of healthcare and medicine.

Importance of Studying Bacteriology

Bacteriology is a crucial field of study within the realm of microbiology, focusing on the identification, classification, and understanding of bacteria. In the book "Bacteriology Uncovered: The Secrets of Microbial World Revealed," the importance of studying bacteriology is emphasized as a fundamental aspect of modern medicine and public health. By delving into the world of bacteria, researchers and healthcare professionals can gain valuable insights into the mechanisms of infectious diseases, antibiotic resistance, and the role of microbiota in maintaining human health.

One of the primary reasons why studying bacteriology is essential is its significant impact on public health. Bacteria are ubiquitous in the environment and can cause a wide range of infectious diseases in humans, ranging from minor skin infections to life-threatening conditions such as pneumonia and sepsis. By understanding the characteristics and behavior of various bacterial species, scientists can develop effective strategies for preventing, diagnosing, and treating these infections, ultimately improving public health outcomes.

Furthermore, bacteriology plays a crucial role in the field of antibiotic resistance, a growing concern in modern healthcare. As bacteria evolve and develop resistance to commonly used antibiotics, it becomes increasingly challenging to treat bacterial infections effectively. By studying bacteriology, researchers can gain a deeper understanding of the mechanisms of antibiotic resistance and develop new drugs and treatment approaches to combat this global threat.

In addition to its impact on infectious diseases and antibiotic resistance, bacteriology also sheds light on the complex interactions between bacteria and the human body. The human microbiota, comprised of trillions of bacteria living in and on our bodies, plays a vital role in maintaining our health and well-being. By studying bacteriology, researchers can uncover the intricate relationships between the microbiota and various aspects of human health, such as digestion, immune function, and mental health.

Overall, the study of bacteriology is essential for advancing our understanding of bacteria and their impact on human health. By delving into the secrets of the microbial world, researchers and healthcare professionals can develop innovative solutions to combat infectious diseases, antibiotic resistance, and other public health challenges. "Bacteriology Uncovered" serves as a valuable resource for those interested in exploring the fascinating field of bacteriology and its implications for modern medicine and public health.

Chapter 2

The Microbial World

Classification of Microorganisms

Classification of microorganisms is essential in the field of bacteriology as it helps scientists and researchers understand the diversity and characteristics of different types of bacteria.

Microorganisms are classified based on various criteria such as shape, size, biochemical characteristics, and genetic makeup. By categorizing microorganisms into different groups, scientists can study their behavior, interactions with other organisms, and potential impact on human health.

One of the most common ways to classify microorganisms is based on their shape. Bacteria can be classified as cocci (spherical), bacilli (rod-shaped), or spirilla (spiral-shaped). Each shape has unique characteristics that can help scientists identify and differentiate between different types of bacteria. For example, cocci bacteria are often found in clusters or chains, while bacilli bacteria are typically found singly or in pairs.

Another important criterion for classifying microorganisms is based on their biochemical characteristics. Bacteria can be classified based on their ability to utilize certain nutrients, produce specific enzymes, or survive in different environments. This classification helps scientists understand how bacteria interact with their surroundings and how they contribute to various biological processes.

Genetic makeup is also a crucial factor in classifying microorganisms. Bacteria can be classified based on their genetic material, such as DNA or RNA, and the presence of specific genes that determine their characteristics. By studying the genetic makeup of bacteria, scientists can identify similarities and differences between different types of bacteria and trace their evolutionary history.

Overall, classification of microorganisms is a fundamental aspect of bacteriology that helps scientists organize and understand the vast diversity of bacteria in the microbial world. By categorizing microorganisms based on their shape, biochemical characteristics, and genetic makeup, scientists can gain insights into their behavior, interactions, and potential impact on human health. This classification system is essential for furthering our knowledge of bacteria and developing strategies for combating infectious diseases and promoting human health.

Characteristics of Bacteria

Bacteria are single-celled organisms that are found virtually everywhere on Earth. They can be found in soil, water, air, and even inside the human body. Despite their small size, bacteria play a crucial role in various ecosystems and are essential for many biological processes. In this subchapter, we will discuss the characteristics of bacteria that make them unique and important in the microbial world.

One of the key characteristics of bacteria is their small size. Most bacteria are microscopic, ranging in size from 0.5 to 5 micrometers in length. This small size allows bacteria to rapidly multiply and adapt to different environments. Additionally, their size enables them to easily pass through various barriers, such as cell membranes, making them versatile and resilient organisms.

Bacteria are also known for their diverse shapes and structures. While some bacteria are spherical (cocci), others are rod-shaped (bacilli) or spiral-shaped (spirilla). These different shapes allow bacteria to perform various functions, such as moving through their environment or attaching to surfaces. The diversity in bacterial shapes also reflects their ability to thrive in different habitats and conditions.

Another important characteristic of bacteria is their ability to reproduce quickly. Under optimal conditions, bacteria can divide every 20 minutes, leading to exponential growth. This rapid reproduction allows bacteria to colonize new environments and outcompete other organisms for resources. However, this fast growth rate can also lead to the spread of harmful bacteria and infectious diseases.

Bacteria are also known for their ability to metabolize a wide range of nutrients. Unlike humans and other animals, bacteria can use a variety of organic and inorganic compounds as a source of energy. This metabolic flexibility allows bacteria to survive in diverse environments, from extreme temperatures to acidic conditions. Additionally, some bacteria can produce enzymes that are beneficial for industrial processes, such as fermentation and waste treatment.

Overall, the characteristics of bacteria make them fascinating and essential organisms in the microbial world. Their small size, diverse shapes, rapid reproduction, and metabolic flexibility enable them to thrive in a wide range of environments and perform vital functions in various ecosystems. By understanding the unique characteristics of bacteria, we can gain insights into their role in nature, as well as their potential applications in medicine, agriculture, and biotechnology.

Differences Between Bacteria and Other Microorganisms

Bacteria are a type of microorganism that play a crucial role in the microbial world. However, they are not the only type of microorganism out there. In this subchapter, we will explore the differences between bacteria and other microorganisms to gain a better understanding of the diversity of the microbial world.

One of the key differences between bacteria and other microorganisms is their cellular structure. Bacteria are prokaryotic organisms, meaning they do not have a nucleus or membrane-bound organelles. In contrast, other microorganisms such as fungi, protists, and algae are eukaryotic organisms, which means they have a nucleus and membrane-bound organelles. This difference in cellular structure has significant implications for the way these organisms function and interact with their environment.

Another difference between bacteria and other microorganisms is their mode of reproduction. Bacteria reproduce asexually through binary fission, where a single cell divides into two identical daughter cells. On the other hand, other microorganisms reproduce sexually or asexually through processes such as budding, fragmentation, or spore formation. This difference in reproductive strategies contributes to the diversity and adaptability of microorganisms in various environments.

Additionally, bacteria and other microorganisms have different nutritional requirements. Bacteria are capable of utilizing a wide range of nutrients, including sugars, proteins, and fats. Some bacteria are even capable of photosynthesis, using sunlight as a source of energy. In contrast, other microorganisms have specific nutritional requirements and may rely on organic matter or other organisms for their survival. Understanding these differences in nutritional requirements is essential for studying the ecology and interactions of microorganisms in different environments.

Furthermore, bacteria and other microorganisms differ in their ecological roles and impacts on human health. While some bacteria are beneficial and play important roles in processes such as nutrient cycling and food production, others can be pathogenic and cause diseases in humans, animals, and plants. Similarly, other microorganisms such as fungi and protists can have both beneficial and harmful effects on their environment and the organisms they interact with. By understanding these differences, we can better appreciate the complexity and diversity of the microbial world and develop strategies for managing and controlling microbial populations for the benefit of human health and the environment.

Chapter 3

Bacterial Structure and Function

Cell Structure of Bacteria

Bacteria are single-celled microorganisms that play a crucial role in various ecosystems, including the human body. Understanding the cell structure of bacteria is essential for studying their functions, interactions, and potential threats to human health. In this subchapter, we will delve into the intricate details of the cell structure of bacteria and explore how these microscopic organisms are organized.

The cell structure of bacteria is relatively simple compared to eukaryotic cells but is still highly efficient in carrying out essential functions. Bacterial cells are typically composed of a cell wall, cell membrane, cytoplasm, ribosomes, and genetic material in the form of DNA. The cell wall of bacteria provides structural support and protection, while the cell membrane regulates the passage of nutrients and waste products in and out of the cell.

One of the defining features of bacterial cells is the lack of a true nucleus. Instead, the genetic material of bacteria is typically organized in a single circular chromosome located in the nucleoid region of the cell. Bacteria may also contain plasmids, small circular pieces of DNA that can confer various traits such as antibiotic resistance. The presence of plasmids allows bacteria to adapt rapidly to changing environmental conditions.

In addition to their genetic material, bacterial cells also contain ribosomes, which are responsible for protein synthesis. These ribosomes are smaller than those found in eukaryotic cells, making them a target for certain antibiotics that specifically inhibit bacterial protein synthesis. The cytoplasm of bacterial cells contains various enzymes and molecules necessary for metabolism, growth, and reproduction.

Overall, the cell structure of bacteria is highly specialized for their unique lifestyle as single-celled organisms. By understanding the organization and functions of bacterial cells, researchers can develop new strategies for combating bacterial infections, developing biotechnological applications, and exploring the diverse roles that bacteria play in the natural world. This subchapter provides a comprehensive overview of the cell structure of bacteria, shedding light on the secrets of the microbial world and the fascinating biology of these tiny but mighty organisms.

Functions of Bacterial Components

In the subchapter titled "Functions of Bacterial Components" in the book "Bacteriology Uncovered: The Secrets of Microbial World Revealed," we delve into the intricate workings of bacterial components and their vital roles in the microbial world. Bacteria are single-celled organisms that possess a variety of structures and components that enable them to thrive in diverse environments. Understanding the functions of these components is essential for comprehending the behavior and characteristics of bacteria.

One of the key components of bacteria is the cell wall, which provides structural support and protection for the cell. The cell wall is composed of peptidoglycan, a unique molecule that is not found in eukaryotic cells. The cell wall helps bacteria maintain their shape and protects them from environmental stressors such as osmotic pressure. Additionally, the cell wall plays a crucial role in bacterial pathogenicity, as it can interact with host cells and contribute to the virulence of pathogenic bacteria.

Another important bacterial component is the plasma membrane, which surrounds the cytoplasm and regulates the passage of molecules in and out of the cell. The plasma membrane is made up of phospholipids and proteins that help maintain the integrity of the cell and facilitate various cellular processes. It also serves as a barrier that protects the cell from harmful substances in the environment. The plasma membrane is essential for bacterial survival and plays a critical role in nutrient uptake, energy production, and cell communication.

Bacterial flagella are long, whip-like appendages that enable bacteria to move through their environment. Flagella are composed of a protein called flagellin and rotate like propellers to propel the cell forward. Bacterial flagella are important for bacterial motility, allowing bacteria to navigate towards favorable conditions and away from harmful environments. Flagella also play a role in bacterial attachment to surfaces and host cells, facilitating the colonization of new environments.

Pili, also known as fimbriae, are hair-like structures that extend from the surface of bacterial cells. Pili are involved in a variety of functions, including adhesion to surfaces, biofilm formation, and bacterial conjugation. Adhesive pili help bacteria attach to host cells and colonize tissues, while conjugative pili facilitate the transfer of genetic material between bacterial cells. Understanding the functions of pili is crucial for understanding bacterial pathogenicity and the spread of antibiotic resistance genes.

In summary, the functions of bacterial components are diverse and essential for bacterial survival and virulence. The cell wall, plasma membrane, flagella, and pili all play critical roles in bacterial physiology, behavior, and interactions with the environment. By studying these components and their functions, we can gain valuable insights into the biology of bacteria and develop new strategies for combating bacterial infections and diseases.

Cell Division and Reproduction in Bacteria

Cell division and reproduction in bacteria are essential processes that allow these microorganisms to proliferate and colonize various environments. Bacteria reproduce through a process called binary fission, where a single cell divides into two identical daughter cells. This rapid form of reproduction allows bacteria to quickly increase their population size and adapt to changing conditions in their surroundings.

During binary fission, the bacterial cell replicates its genetic material and then divides into two daughter cells, each containing a copy of the original DNA. This process is highly efficient and allows bacteria to reproduce at a rapid rate, with some species dividing every 20 minutes under optimal conditions. The ability of bacteria to reproduce quickly is one of the reasons why they are so successful at colonizing diverse habitats, from soil and water to the human body.

In addition to binary fission, some bacteria can also reproduce through other mechanisms such as budding or spore formation. Budding involves the formation of a small outgrowth on the parent cell, which eventually detaches to become a new individual. Spore formation, on the other hand, allows bacteria to survive in harsh environmental conditions by forming dormant, resistant spores that can germinate when conditions become favorable again.

Understanding the mechanisms of cell division and reproduction in bacteria is crucial for studying their growth and behavior in different environments. By studying how bacteria reproduce, scientists can gain insights into how these microorganisms adapt to changing conditions, develop resistance to antibiotics, and cause infections in humans and animals. This knowledge is essential for the development of new strategies to control bacterial growth and prevent the spread of infectious diseases.

In conclusion, cell division and reproduction are fundamental processes in the life cycle of bacteria. Through binary fission, budding, and spore formation, bacteria can rapidly increase their population size and adapt to diverse environments. By studying the mechanisms of bacterial reproduction, scientists can uncover the secrets of the microbial world and develop new approaches to combat bacterial infections and promote human health.

Chapter 4

Bacterial Growth and Metabolism

Factors Affecting Bacterial Growth

Bacterial growth is a complex process that can be influenced by a variety of factors. Understanding these factors is crucial in both clinical and research settings, as they can impact the effectiveness of treatments and the spread of infectious diseases. In this subchapter, we will explore some of the key factors that can affect bacterial growth.

One of the most important factors affecting bacterial growth is the availability of nutrients. Bacteria require a range of nutrients, including carbon, nitrogen, and phosphorus, to grow and reproduce. The availability of these nutrients in the environment can greatly impact the growth rate of bacteria. For example, bacteria that are able to access a plentiful food source will be able to grow and reproduce more quickly than those that are limited by nutrient availability.

Another important factor that can affect bacterial growth is temperature. Bacteria are classified into different groups based on their temperature preferences, including psychrophiles (cold-loving bacteria), mesophiles (moderate-temperature bacteria), and thermophiles (heat-loving bacteria). The temperature at which a bacterium thrives can impact its growth rate, with most bacteria growing best at moderate temperatures around 37°C.

pH levels can also play a significant role in bacterial growth. Bacteria have specific pH ranges at which they can survive and reproduce, with most bacteria preferring neutral pH levels around 7. However, some bacteria are able to thrive in acidic or alkaline environments, such as the acid-loving bacteria that cause stomach ulcers. Changes in pH levels can impact bacterial growth by disrupting cellular processes and enzyme activity.

Oxygen availability is another factor that can affect bacterial growth. While some bacteria require oxygen to grow (obligate aerobes), others are killed by exposure to oxygen (obligate anaerobes). Facultative anaerobes are able to grow in the presence or absence of oxygen. The availability of oxygen in the environment can impact the growth rate and metabolism of bacteria, with oxygen-rich environments typically supporting faster growth rates.

In conclusion, bacterial growth is a complex process that can be influenced by a variety of factors, including nutrient availability, temperature, pH levels, and oxygen availability. Understanding these factors is essential for researchers and clinicians working in the field of bacteriology, as they can impact the effectiveness of treatments and the spread of infectious diseases. By studying the factors that affect bacterial growth, we can gain valuable insights into how bacteria interact with their environment and develop strategies to control their growth and spread.

Nutritional Requirements of Bacteria

Bacteria are incredibly diverse organisms that play a crucial role in various ecosystems, including the human body. To thrive and carry out their functions effectively, bacteria have specific nutritional requirements that must be met. Understanding these requirements is essential for studying and manipulating bacterial populations in different environments. In this subchapter, we will delve into the nutritional requirements of bacteria and explore how they obtain the nutrients necessary for their survival and growth.

One of the key nutritional requirements of bacteria is a source of carbon. Carbon is essential for building the organic molecules that make up the bacterial cell. Bacteria can be classified based on their carbon source as either autotrophs, which can synthesize their own organic molecules from inorganic carbon sources, or heterotrophs, which rely on organic compounds as their carbon source. Understanding the carbon requirements of bacteria is crucial for designing growth media that support the growth of specific bacterial species in the laboratory.

In addition to carbon, bacteria require a source of energy to fuel their metabolic processes. Bacteria can be classified based on their energy source as either phototrophs, which use light energy to drive their metabolic reactions, or chemotrophs, which rely on chemical energy sources. Chemotrophs can further be divided into chemoautotrophs, which use inorganic compounds as their energy source, and chemoheterotrophs, which obtain energy from organic compounds. By understanding the energy requirements of bacteria, researchers can manipulate bacterial metabolism and growth in various environments.

Bacteria also require a range of essential nutrients, such as nitrogen, phosphorus, sulfur, and trace elements, to support their growth and metabolic processes. These nutrients are essential for building proteins, nucleic acids, and other vital molecules in the bacterial cell. Understanding the specific nutrient requirements of different bacterial species is crucial for designing growth media that promote their growth and studying their metabolic capabilities. By providing the necessary nutrients, researchers can cultivate specific bacterial species in the laboratory and investigate their biological properties.

Overall, the nutritional requirements of bacteria are diverse and complex, reflecting the incredible adaptability of these microorganisms to different environments. By understanding the carbon, energy, and nutrient requirements of bacteria, researchers can manipulate bacterial populations in various ecosystems and study their roles in different processes. Investigating the nutritional requirements of bacteria is a fundamental aspect of bacteriology that sheds light on the secrets of the microbial world and opens up new possibilities for understanding and harnessing the power of these tiny organisms.

Metabolic Pathways in Bacteria

Metabolic pathways in bacteria play a crucial role in the survival and growth of these microorganisms. These pathways involve a series of chemical reactions that help bacteria obtain energy, synthesize essential biomolecules, and adapt to their environment. Understanding these metabolic pathways is key to studying bacterial physiology and developing new strategies to combat bacterial infections.

One of the most important metabolic pathways in bacteria is glycolysis. This pathway involves the breakdown of glucose into pyruvate, which can then be used to generate ATP, the energy currency of the cell. Glycolysis is a central metabolic pathway that is conserved across all living organisms, including bacteria. By studying glycolysis in bacteria, researchers can gain insights into how these microorganisms obtain energy and nutrients in different environments.

Another key metabolic pathway in bacteria is the tricarboxylic acid (TCA) cycle, also known as the citric acid cycle. This pathway is involved in the generation of ATP through the oxidation of acetyl-CoA, a molecule derived from the breakdown of carbohydrates, fats, and proteins. The TCA cycle is essential for the production of energy in bacteria and is also important for the synthesis of precursors for the biosynthesis of amino acids, nucleotides, and other biomolecules.

In addition to glycolysis and the TCA cycle, bacteria also have unique metabolic pathways that allow them to utilize a wide range of carbon and energy sources. For example, some bacteria can ferment sugars to produce energy, while others can use alternative carbon sources such as organic acids, alcohols, and gases. By studying these diverse metabolic pathways, researchers can gain insights into the metabolic flexibility of bacteria and their ability to survive in various environments.

Overall, understanding the metabolic pathways in bacteria is essential for unraveling the secrets of the microbial world. By studying how bacteria obtain energy, synthesize biomolecules, and adapt to different environments, researchers can develop new strategies to combat bacterial infections and harness the potential of these microorganisms for biotechnological applications. The study of metabolic pathways in bacteria continues to be a fascinating and important field of research in bacteriology.

Chapter 5

Bacterial Genetics

DNA Replication in Bacteria

DNA replication is a fundamental process in all living organisms, including bacteria. In bacteria, DNA replication is a highly precise and efficient process that ensures the faithful transmission of genetic information from one generation to the next. Understanding the mechanisms of DNA replication in bacteria is essential for unraveling the mysteries of microbial genetics and evolution.

In bacteria, DNA replication is initiated at a specific site on the chromosome called the origin of replication. This site is recognized by a complex of proteins that unwind the DNA double helix and create a replication fork, where DNA synthesis can begin. The process of DNA replication in bacteria is highly coordinated, with multiple enzymes and proteins working together to ensure accuracy and efficiency.

One of the key enzymes involved in DNA replication in bacteria is DNA polymerase. This enzyme is responsible for synthesizing a new DNA strand by adding nucleotides one by one to the growing chain. DNA polymerase is highly accurate, with a proofreading function that helps to correct errors in the DNA sequence. In addition to DNA polymerase, other enzymes such as helicase, primase, and ligase are also involved in the process of DNA replication in bacteria.

During DNA replication in bacteria, the two strands of the DNA double helix are unwound and separated by helicase enzymes. Primase then synthesizes a short RNA primer that serves as a starting point for DNA synthesis. DNA polymerase then extends the RNA primer by adding complementary nucleotides to the growing DNA strand. Finally, ligase seals any breaks in the DNA backbone to produce two identical daughter DNA molecules.

Overall, DNA replication in bacteria is a highly regulated and efficient process that ensures the accurate transmission of genetic information from one generation to the next. Understanding the mechanisms of DNA replication in bacteria is essential for studying microbial genetics and evolution, as well as for developing new strategies for combating bacterial infections. By unraveling the secrets of DNA replication in bacteria, scientists can unlock the mysteries of the microbial world and pave the way for new discoveries in the field of bacteriology.

Gene Expression and Regulation in Bacteria

Gene expression and regulation in bacteria play a crucial role in the survival and adaptation of these microorganisms. Bacteria must tightly regulate the expression of their genes in order to respond to changes in their environment, such as nutrient availability, temperature, and pH. This regulation allows bacteria to conserve energy and resources by only producing the proteins needed at any given time.

One of the key mechanisms of gene expression regulation in bacteria is through the use of transcription factors. These proteins bind to specific DNA sequences, known as promoters, and either activate or repress the transcription of nearby genes. By controlling the rate at which genes are transcribed into messenger RNA (mRNA), bacteria can quickly respond to changes in their environment.

In addition to transcription factors, bacteria also utilize small regulatory RNA molecules to control gene expression. These small RNAs can base-pair with mRNA molecules, leading to either their degradation or inhibition of translation. This post-transcriptional regulation allows bacteria to fine-tune the expression of specific genes without the need for extensive changes in transcription.

Another important aspect of gene expression regulation in bacteria is the use of operons. Operons are clusters of genes that are transcribed together as a single mRNA molecule. This allows bacteria to coordinate the expression of multiple genes involved in a common pathway or cellular process. By regulating the expression of entire operons, bacteria can quickly respond to environmental changes and ensure the proper functioning of essential cellular processes.

Overall, the regulation of gene expression in bacteria is a complex and tightly controlled process that allows these microorganisms to adapt to changing environmental conditions. By understanding the mechanisms underlying gene expression regulation in bacteria, researchers can gain valuable insights into how these organisms survive and thrive in diverse environments.

Horizontal Gene Transfer in Bacteria

Horizontal gene transfer in bacteria is a fascinating phenomenon that plays a significant role in the evolution and adaptation of microbial populations. Unlike vertical gene transfer, which occurs through reproduction, horizontal gene transfer involves the transfer of genetic material between different bacterial cells. This process allows bacteria to acquire new genes that can confer advantageous traits, such as antibiotic resistance or the ability to metabolize new nutrients.

There are several mechanisms by which horizontal gene transfer can occur in bacteria. One common method is through the process of conjugation, where a plasmid containing the desired gene is transferred from one bacterial cell to another through direct contact. Another mechanism is transformation, where bacteria can take up free DNA from their environment and incorporate it into their own genetic material. Lastly, transduction involves the transfer of genetic material through a viral vector known as a bacteriophage.

Horizontal gene transfer has important implications for bacterial pathogenesis and antibiotic resistance. The exchange of genes between different bacterial species can lead to the spread of virulence factors, making pathogenic bacteria more potent. Additionally, the transfer of antibiotic resistance genes can result in the emergence of multidrug-resistant strains, posing a significant challenge to the treatment of bacterial infections.

Understanding the mechanisms and implications of horizontal gene transfer in bacteria is crucial for the development of effective strategies to combat antibiotic resistance and prevent the spread of virulence factors. By studying the genetic exchange between bacterial populations, researchers can gain insights into the evolutionary processes that drive microbial diversity and adaptability. This knowledge can inform the design of novel antimicrobial therapies and help us stay one step ahead of bacterial pathogens.

A Primer for Healthcare Professionals

Chapter 6

Bacterial Pathogenesis

Mechanisms of Bacterial Pathogenicity

In the subchapter "Mechanisms of Bacterial Pathogenicity" we delve into the intricate ways in which bacteria are able to cause disease within the human body. Understanding these mechanisms is crucial in developing effective strategies for preventing and treating bacterial infections.

One of the key mechanisms of bacterial pathogenicity is the ability of bacteria to adhere to and invade host tissues. Bacteria possess specialized structures such as pili and adhesins that enable them to attach to specific receptors on host cells. Once attached, bacteria can then invade host cells, allowing them to evade the immune system and cause further damage.

Another important mechanism of bacterial pathogenicity is the production of toxins. Toxins are molecules that are released by bacteria and can cause damage to host cells and tissues. Some bacteria produce toxins that directly damage host cells, while others produce toxins that interfere with the host immune response. Understanding the role of toxins in bacterial pathogenicity is essential for developing treatments that target these molecules.

Bacteria are also able to evade the host immune system through a variety of mechanisms. Some bacteria have evolved ways to hide from the immune system by modifying their surface structures or by producing molecules that inhibit the immune response. Others are able to survive and replicate within host cells, making them difficult for the immune system to detect and eliminate.

In addition to evading the immune system, bacteria can also manipulate host cell signaling pathways to their advantage. By interfering with host cell signaling, bacteria can promote their own survival and replication within the host. Understanding how bacteria manipulate host cell signaling pathways is important for developing targeted therapies that disrupt these interactions.

Overall, the mechanisms of bacterial pathogenicity are complex and varied. By studying these mechanisms in detail, researchers can gain insights into how bacteria cause disease and develop new strategies for preventing and treating bacterial infections. This knowledge is crucial for advancing our understanding of bacterial diseases and improving patient outcomes.

Host-Pathogen Interactions

Host-pathogen interactions are a critical aspect of bacteriology that plays a significant role in the development and progression of infectious diseases. In this subchapter, we will explore the intricate relationship between hosts and pathogens, highlighting the various mechanisms by which bacteria interact with their host organisms.

One of the key aspects of host-pathogen interactions is the ability of bacteria to evade the host immune system. Bacteria have evolved a wide range of strategies to avoid detection and destruction by the immune system, including the production of virulence factors that enable them to invade host cells and tissues. Understanding these mechanisms is crucial for developing effective treatments and vaccines for bacterial infections.

Another important aspect of host-pathogen interactions is the role of the host microbiome in modulating immune responses to bacterial pathogens. The microbiome, which consists of trillions of bacteria that live in and on our bodies, plays a crucial role in regulating immune function and maintaining homeostasis. Disruption of the microbiome can result in dysregulation of the immune system, making the host more susceptible to bacterial infections.

Furthermore, host-pathogen interactions can lead to the development of chronic infections that are difficult to treat. Bacteria have evolved complex mechanisms to survive and persist in the host environment, including the formation of biofilms that protect them from immune attack and antibiotic therapy. Understanding these mechanisms is essential for developing new therapies to combat chronic bacterial infections.

In conclusion, host-pathogen interactions are a fascinating and complex area of study in bacteriology that has far-reaching implications for human health. By unraveling the secrets of how bacteria interact with their hosts, we can develop new strategies to prevent and treat infectious diseases, ultimately improving the lives of millions of people around the world.

Immune Response to Bacterial Infections

The immune response to bacterial infections is a complex and highly coordinated process that involves multiple components of the immune system working together to protect the body from harmful bacteria. When a bacterial infection occurs, the body's first line of defense is the innate immune system, which includes physical barriers like the skin and mucous membranes, as well as cells like macrophages and neutrophils that can engulf and destroy bacteria.

Once the innate immune system has detected the presence of bacteria, it triggers an inflammatory response that helps to recruit more immune cells to the site of infection. This inflammatory response is characterized by redness, swelling, heat, and pain, which are all signs that the immune system is working to fight off the infection. In addition to recruiting immune cells, the inflammatory response also helps to create an environment that is hostile to bacteria, making it more difficult for them to survive and replicate.

One of the key components of the adaptive immune response to bacterial infections is the production of antibodies. Antibodies are proteins that can specifically recognize and bind to the surface of bacteria, marking them for destruction by other immune cells. In addition to antibodies, the adaptive immune response also involves the activation of T cells, which can directly kill infected cells and help to regulate the overall immune response.

In some cases, the immune response to a bacterial infection can become dysregulated, leading to conditions like sepsis or septic shock. These conditions occur when the immune system overreacts to the presence of bacteria, causing widespread inflammation and damage to tissues throughout the body. In severe cases, sepsis and septic shock can be life-threatening and require immediate medical intervention to prevent organ failure and death.

Overall, the immune response to bacterial infections is a remarkable example of the body's ability to defend itself against harmful pathogens. By understanding the different components of the immune response and how they work together to fight off bacterial infections, researchers and healthcare providers can develop new strategies for treating and preventing these common and sometimes deadly infections.

Chapter 7

Antibiotics and Bacterial Resistance

Overview of Antibiotics

Antibiotics are powerful medications that are used to treat bacterial infections by either killing the bacteria or preventing their growth. They have revolutionized the field of medicine and have played a crucial role in saving countless lives. In this subchapter, we will delve into the fascinating world of antibiotics and explore their mechanisms of action, types, and the importance of proper use.

The discovery of antibiotics is attributed to Sir Alexander Fleming, who accidentally stumbled upon penicillin in 1928. This chance discovery paved the way for the development of numerous other antibiotics that have been instrumental in combating various bacterial infections. Antibiotics work by targeting specific components of bacterial cells, such as cell walls or protein synthesis machinery, while leaving human cells unharmed.

There are several classes of antibiotics, each with its own mechanism of action and spectrum of activity. Some common classes of antibiotics include penicillins, cephalosporins, tetracyclines, macrolides, and fluoroquinolones. Each class targets different types of bacteria, so it is important for healthcare providers to choose the right antibiotic based on the specific infection being treated.

It is crucial to use antibiotics judiciously to prevent the development of antibiotic resistance. Overuse or misuse of antibiotics can lead to the emergence of drug-resistant bacteria, making infections more difficult to treat. Healthcare providers should prescribe antibiotics only when necessary, and patients should always follow the prescribed dosage and duration of treatment to ensure the effectiveness of the medication.

In conclusion, antibiotics have been a game-changer in the field of medicine and have saved countless lives. Understanding the mechanisms of action and proper use of antibiotics is essential for healthcare providers and patients alike. By using antibiotics responsibly and in accordance with guidelines, we can help preserve their efficacy and continue to benefit from their life-saving properties.

Mechanisms of Antibiotic Action

Antibiotics are powerful drugs that are used to treat bacterial infections in both humans and animals. They work by targeting specific mechanisms within bacteria that are essential for their survival and growth. Understanding these mechanisms of antibiotic action is crucial for the development of new antibiotics and the prevention of antibiotic resistance.

One of the most common mechanisms of antibiotic action is the inhibition of bacterial cell wall synthesis. Many antibiotics, such as penicillin and cephalosporins, work by interfering with the formation of peptidoglycan, a key component of the bacterial cell wall. Without a fully functional cell wall, bacteria are unable to maintain their shape and integrity, leading to their eventual death.

Another important mechanism of antibiotic action is the inhibition of protein synthesis. Antibiotics like tetracyclines and macrolides target the ribosomes of bacterial cells, preventing them from producing essential proteins. Without these proteins, bacteria are unable to carry out vital cellular processes, ultimately leading to their demise.

Some antibiotics work by disrupting bacterial DNA replication and repair. Fluoroquinolones, for example, target enzymes that are involved in these processes, leading to the accumulation of DNA damage within bacterial cells. This damage is toxic to the bacteria and eventually results in cell death.

In addition to these mechanisms, some antibiotics disrupt bacterial cell membranes, interfere with metabolic pathways, or inhibit essential enzymes. By targeting different aspects of bacterial physiology, antibiotics are able to effectively kill or inhibit the growth of a wide range of bacterial species. Understanding these mechanisms of action is essential for the rational design of new antibiotics and the development of strategies to combat antibiotic resistance.

Bacterial Resistance to Antibiotics

Bacterial resistance to antibiotics is a growing concern in the field of bacteriology. Antibiotics are powerful medications that are designed to kill or inhibit the growth of bacteria. However, over time, bacteria can develop resistance to these drugs, making them less effective in treating infections. This phenomenon is known as antibiotic resistance.

There are several factors that contribute to bacterial resistance to antibiotics. One of the main reasons is the overuse and misuse of antibiotics. When antibiotics are used unnecessarily or not taken as prescribed, bacteria have the opportunity to develop resistance. In addition, the widespread use of antibiotics in agriculture and livestock can also contribute to the development of resistant bacteria.

Another factor that contributes to bacterial resistance is the ability of bacteria to adapt and evolve. Bacteria have the ability to mutate and acquire resistance genes through mechanisms such as horizontal gene transfer. This allows them to develop resistance to antibiotics and pass this resistance on to other bacteria.

In order to combat bacterial resistance to antibiotics, it is important for healthcare providers to prescribe antibiotics judiciously and for patients to take them as directed. Additionally, researchers are working to develop new antibiotics and alternative treatments to combat resistant bacteria. It is also important for policymakers to implement regulations to limit the use of antibiotics in agriculture and livestock to help prevent the spread of resistance.

Overall, understanding bacterial resistance to antibiotics is crucial in the field of bacteriology. By studying the mechanisms behind resistance and working to develop new treatment options, researchers can help combat this growing problem and ensure that antibiotics remain effective in treating bacterial infections.

Chapter 8

Industrial and Environmental Applications of Bacteriology

Role of Bacteria in Industry

Bacteria play a crucial role in various industries, from food production to pharmaceuticals. In the food industry, bacteria are used to ferment dairy products like yogurt and cheese, as well as to produce beverages such as beer and wine. The process of fermentation involves the conversion of sugars into alcohol and carbon dioxide by bacteria, which gives these products their unique flavors and textures. Bacteria are also used in the production of vitamins, enzymes, and antibiotics, which are essential components of many pharmaceutical products.

In the agricultural industry, bacteria are used to promote plant growth and protect crops from diseases. Some bacteria form symbiotic relationships with plants, providing them with essential nutrients like nitrogen in exchange for sugars produced by the plants through photosynthesis. Other bacteria produce antibiotics that can protect plants from harmful pathogens. By harnessing the power of these beneficial bacteria, farmers can increase crop yields and reduce the need for chemical fertilizers and pesticides.

In the environmental industry, bacteria play a key role in wastewater treatment and bioremediation. Wastewater treatment plants use bacteria to break down organic matter and remove pollutants from water before it is released back into the environment. Bacteria are also used to clean up oil spills and industrial waste by breaking down harmful chemicals and converting them into less toxic substances. By utilizing the natural abilities of bacteria, environmental engineers can help protect our ecosystems and ensure a cleaner, healthier planet for future generations.

In the textile industry, bacteria are used to produce fabrics with unique properties. Certain bacteria can produce cellulose, a natural polymer that is used to create materials like rayon and lyocell. By fermenting bacteria in a controlled environment, manufacturers can create sustainable textiles that are biodegradable and eco-friendly. Bacteria are also used to dye fabrics using natural pigments, reducing the need for synthetic dyes that can harm the environment and human health.

Overall, the role of bacteria in industry is vast and varied, with applications in food production, pharmaceuticals, agriculture, environmental protection, and textiles. By understanding and harnessing the power of these microscopic organisms, we can improve the efficiency and sustainability of various industries while also protecting our health and the environment. Bacteria truly are the unsung heroes of the microbial world, quietly working behind the scenes to keep our industries running smoothly and our planet thriving.

Bioremediation and Environmental Cleanup

Bioremediation is a process that utilizes microorganisms to clean up environmental pollutants. This method is becoming increasingly popular due to its effectiveness and environmentally friendly nature. By harnessing the power of bacteria, fungi, and other microbes, bioremediation can target a wide range of contaminants, including oil spills, heavy metals, and chemical pollutants. This process is not only cost-effective but also sustainable, making it a preferred choice for environmental cleanup efforts.

One of the key benefits of bioremediation is its ability to target specific contaminants without causing harm to the surrounding environment. Unlike traditional cleanup methods, which may involve the use of harsh chemicals or mechanical removal, bioremediation is a natural and non-invasive approach. Microorganisms break down contaminants into harmless byproducts through processes such as biodegradation and biotransformation. This results in a cleaner and healthier environment without the negative side effects often associated with other cleanup methods.

Another advantage of bioremediation is its versatility in addressing different types of pollutants. Microbes have evolved to thrive in a variety of environments, from soil and water to air and even inside living organisms. This adaptability allows bioremediation to be applied in a wide range of settings, including industrial sites, agricultural fields, and even urban areas. By selecting the right combination of microbial species and techniques, bioremediation can effectively target and eliminate specific contaminants, restoring the environment to its natural state.

In addition to its effectiveness in cleaning up pollutants, bioremediation also offers long-term benefits for the environment. Unlike some traditional cleanup methods, which may only provide a temporary solution, bioremediation has the potential to permanently remove contaminants from the environment. By promoting the growth of beneficial microorganisms and enhancing natural processes, bioremediation creates a sustainable solution that can continue to improve environmental quality over time. This makes it a valuable tool for addressing ongoing pollution issues and preventing future environmental damage.

Overall, bioremediation represents a promising approach to environmental cleanup that harnesses the power of microbial communities to address pollution challenges. By utilizing the natural abilities of bacteria, fungi, and other microorganisms, bioremediation offers a safe, efficient, and sustainable method for removing contaminants from the environment. As we continue to explore the secrets of the microbial world, bioremediation will likely play an increasingly important role in protecting our planet and promoting a cleaner, healthier future for all living organisms.

Probiotics and Other Beneficial Bacteria

Probiotics and other beneficial bacteria play a crucial role in maintaining the balance of our gut microbiota and promoting overall health. Probiotics are live bacteria and yeasts that are good for our health, especially our digestive system. They are often referred to as "good" or "friendly" bacteria because they help keep our gut healthy. These beneficial bacteria can be found in various fermented foods like yogurt, kefir, sauerkraut, and kimchi, as well as in dietary supplements.

Probiotics work by restoring the natural balance of bacteria in the gut, which can be disrupted by factors such as poor diet, stress, illness, or the use of antibiotics. When the balance of bacteria in the gut is disturbed, it can lead to digestive issues, such as bloating, gas, and diarrhea. By consuming probiotics, we can help replenish the good bacteria in our gut and support our overall digestive health.

In addition to probiotics, there are other beneficial bacteria that play a role in maintaining our health. For example, certain strains of bacteria have been found to produce vitamins, such as B vitamins and vitamin K, that are essential for our overall health. Other bacteria help break down food in the gut, making it easier for our bodies to absorb nutrients.

Research has also shown that beneficial bacteria can have a positive impact on our immune system. By promoting a healthy balance of bacteria in the gut, probiotics and other beneficial bacteria can help support our immune system's ability to fight off infections and diseases. This is why maintaining a healthy gut microbiota is essential for overall health and well-being.

In conclusion, probiotics and other beneficial bacteria are important for maintaining a healthy gut microbiota and promoting overall health. By consuming foods rich in probiotics and supporting the growth of beneficial bacteria in our gut, we can help support our digestive health, boost our immune system, and improve our overall well-being. Incorporating probiotic-rich foods into our diet and considering supplementation can be beneficial for those looking to optimize their gut health and overall wellness.

Chapter 9

Future Trends in Bacteriology

Emerging Infectious Diseases

In recent years, the emergence of new infectious diseases has become a major concern in the field of bacteriology. These diseases, known as emerging infectious diseases, pose a significant threat to public health due to their potential to spread rapidly and cause widespread illness. Understanding the factors that contribute to the emergence of these diseases is crucial for developing effective prevention and control strategies.

One of the key factors driving the emergence of infectious diseases is the increasing interconnectedness of the modern world. Globalization has led to greater movement of people, animals, and goods across borders, creating new opportunities for pathogens to spread. Climate change is also playing a role in the emergence of infectious diseases, as shifting environmental conditions can create new habitats for disease-carrying organisms.

Another important factor in the emergence of infectious diseases is the evolution of pathogens themselves. Bacteria, viruses, and other microbes have the ability to adapt and evolve rapidly, allowing them to overcome existing immune defenses and develop resistance to antibiotics. This evolutionary arms race between pathogens and hosts is a constant challenge for public health authorities.

In addition to these factors, changes in human behavior and land use can also contribute to the emergence of infectious diseases.

Deforestation, urbanization, and the encroachment of humans into wildlife habitats can bring people into closer contact with disease-carrying animals, increasing the risk of transmission. Poor sanitation and hygiene practices can also create opportunities for pathogens to spread.

Overall, the emergence of infectious diseases is a complex and multifaceted issue that requires a coordinated and interdisciplinary approach to address. By understanding the factors that contribute to the emergence of these diseases, researchers and public health officials can work together to develop strategies for prevention, surveillance, and control. Only through continued vigilance and collaboration can we hope to stay one step ahead of these emerging threats to global health.

Advances in Bacterial Research

Advances in bacterial research have brought about groundbreaking discoveries that have revolutionized our understanding of the microbial world. With the advent of new technologies and methodologies, scientists have been able to delve deeper into the intricate mechanisms of bacteria and uncover their secrets. These advances have paved the way for the development of novel treatments and interventions for bacterial infections, as well as shedding light on the role of bacteria in various ecosystems.

One of the most significant advancements in bacterial research is the use of genomics to study bacterial genomes. With the sequencing of bacterial genomes becoming more accessible and affordable, scientists are now able to analyze the genetic makeup of bacteria with unprecedented detail. This has led to the identification of new bacterial species, as well as the discovery of novel genes and pathways that play crucial roles in bacterial physiology and pathogenesis.

Furthermore, the field of metagenomics has enabled researchers to study the entire microbial community present in a given environment. By analyzing the genetic material of all the microorganisms present in a sample, scientists can gain insights into the diversity and dynamics of bacterial populations. This approach has been instrumental in understanding the role of bacteria in various ecosystems, from the human gut to soil and water environments.

In addition to genomics and metagenomics, advances in imaging technologies have also revolutionized bacterial research. High-resolution microscopy techniques now allow scientists to visualize bacteria at the nanometer scale, revealing their intricate structures and interactions with other organisms. This has provided valuable insights into the behavior of bacteria in their natural habitats, as well as their interactions with host cells during infection.

Overall, the advances in bacterial research have expanded our knowledge of the microbial world and opened up new avenues for the development of treatments and interventions for bacterial infections. By harnessing the power of genomics, metagenomics, and imaging technologies, scientists are uncovering the secrets of bacteria and paving the way for a deeper understanding of their role in health and disease.

Ethical Considerations in Bacteriology

Ethical considerations play a crucial role in the field of bacteriology, as researchers and scientists must adhere to strict guidelines to ensure the safety and well-being of both humans and the environment. When conducting research involving bacteria, it is essential to consider the potential risks and benefits of the study, as well as the ethical implications of the research methods used.

One of the primary ethical considerations in bacteriology is the use of human and animal subjects in research. Researchers must obtain informed consent from participants before conducting any experiments involving bacteria, ensuring that they are aware of the potential risks and benefits of the study. Additionally, researchers must follow strict guidelines for the humane treatment of animals used in bacteriology research, minimizing any potential harm or suffering inflicted on the animals.

Another important ethical consideration in bacteriology is the proper disposal of biohazardous materials. Bacteria can pose a significant risk to human health and the environment if not handled and disposed of properly. Researchers must follow strict protocols for the storage, transportation, and disposal of biohazardous materials to prevent the spread of harmful bacteria and protect public health.

In addition to the ethical considerations surrounding research and experimentation, bacteriologists must also consider the potential implications of their work on society as a whole. The development of new antibiotics and vaccines, for example, can have far-reaching implications for public health and could potentially save countless lives. However, researchers must also consider the potential misuse of their work, such as the development of antibiotic-resistant bacteria or bioterrorism.

Overall, ethical considerations are a crucial aspect of bacteriology research, ensuring that scientists and researchers conduct their work responsibly and ethically. By following strict guidelines and protocols, bacteriologists can help to advance our understanding of bacteria while also protecting public health and the environment.

Chapter 10

Conclusion

Recap of Key Concepts

In this subchapter, we will recap some of the key concepts that have been covered in our exploration of bacteriology. Understanding the microbial world is crucial for medical professionals, researchers, and anyone interested in the fascinating world of bacteria. By revisiting these key concepts, we can solidify our understanding of the subject and continue to uncover the secrets of the microbial world.

One of the fundamental concepts in bacteriology is the classification of bacteria. Bacteria are classified based on their shape, structure, and the way they obtain energy. Understanding the different types of bacteria is essential for identifying and treating bacterial infections. From cocci to spirilla, each type of bacteria has unique characteristics that play a role in their behavior and impact on human health.

Another important concept in bacteriology is antibiotic resistance. Over the years, bacteria have developed resistance to many common antibiotics, making it increasingly difficult to treat infections. Understanding how bacteria develop resistance and the importance of responsible antibiotic use is crucial in the fight against antibiotic-resistant bacteria.

The role of bacteria in human health is also a key concept in bacteriology. While some bacteria can cause harmful infections, many bacteria are essential for human health. The human gut, for example, is home to trillions of bacteria that play a crucial role in digestion, immune function, and overall health. Understanding the symbiotic relationship between humans and bacteria is essential for maintaining optimal health.

Lastly, the concept of bacterial evolution is an important aspect of bacteriology. Bacteria have been evolving for billions of years, adapting to new environments and developing new survival strategies. Understanding how bacteria evolve can provide valuable insights into how to combat antibiotic resistance and develop new treatments for bacterial infections. By studying the evolution of bacteria, we can gain a deeper understanding of the microbial world and uncover new secrets waiting to be revealed.

Implications of Bacteriology in Healthcare

Bacteriology plays a crucial role in healthcare, shaping the way we understand and treat various diseases. By studying bacteria and their interactions with the human body, healthcare professionals are better equipped to diagnose and treat infections. Understanding the implications of bacteriology in healthcare is essential for improving patient outcomes and preventing the spread of infectious diseases.

One of the key implications of bacteriology in healthcare is the development of antibiotics. Antibiotics are drugs that target and kill bacteria, helping to treat bacterial infections. Bacteriology has played a vital role in the discovery and development of antibiotics, revolutionizing the field of medicine. However, the overuse and misuse of antibiotics have led to the rise of antibiotic-resistant bacteria, posing a significant challenge to healthcare professionals worldwide.

Another important implication of bacteriology in healthcare is the understanding of how bacteria contribute to the development of chronic diseases. Research has shown that certain bacteria in the gut microbiome can influence our overall health, playing a role in conditions such as obesity, diabetes, and inflammatory bowel disease. By studying the interactions between bacteria and the human body, healthcare professionals can develop targeted treatments to improve patient outcomes.

Bacteriology also plays a crucial role in infection control and prevention in healthcare settings. Understanding how bacteria spread and cause infections is essential for implementing effective strategies to prevent the transmission of disease. Healthcare professionals use bacteriological techniques to identify and track the spread of infectious bacteria, allowing for quick and targeted interventions to stop outbreaks before they escalate.

Overall, the implications of bacteriology in healthcare are vast and far-reaching. From the development of antibiotics to understanding the role of bacteria in chronic diseases, bacteriology continues to shape the way we approach healthcare. By staying informed about the latest research and advancements in bacteriology, healthcare professionals can better protect their patients and improve healthcare outcomes for all.

Future Directions in Bacteriology Research

The field of bacteriology is constantly evolving, with new discoveries and advancements being made every day. As we look towards the future of bacteriology research, there are several key areas that hold promise for further exploration and understanding of the microbial world.

One exciting direction in bacteriology research is the study of microbial communities, also known as the microbiome. By examining the complex interactions between different species of bacteria and how they influence human health, researchers hope to gain a better understanding of the role that these microorganisms play in diseases such as obesity, autoimmune disorders, and even mental health conditions.

Another important area of future research in bacteriology is the development of new antimicrobial agents. With the rise of antibiotic resistance posing a significant threat to global health, scientists are working to discover novel compounds that can effectively combat bacterial infections. This includes exploring natural sources such as plants and marine organisms, as well as harnessing cutting-edge technologies like CRISPR gene editing to target specific bacterial strains.

Advancements in technology are also shaping the future of bacteriology research. High-throughput sequencing techniques have revolutionized our ability to study the genetic makeup of bacteria, allowing researchers to identify new species, track the spread of antibiotic resistance genes, and understand how bacteria adapt to different environments. In addition, the use of artificial intelligence and machine learning algorithms is helping to analyze vast amounts of data and uncover patterns that would be impossible to detect using traditional methods.

Furthermore, the field of bacteriology is increasingly focusing on the role of bacteria in environmental processes. From bioremediation of polluted sites to the production of biofuels and bioplastics, bacteria have the potential to revolutionize industries and contribute to a more sustainable future. By studying how bacteria interact with their surroundings and harnessing their unique abilities, researchers can develop innovative solutions to some of the biggest challenges facing our planet.

In conclusion, the future of bacteriology research is bright and full of exciting possibilities. By exploring the microbiome, developing new antimicrobial agents, leveraging advanced technologies, and harnessing the power of bacteria for environmental applications, scientists are poised to make significant strides in our understanding of the microbial world. As we continue to uncover the secrets of bacteria, we will undoubtedly unlock new opportunities for improving human health, protecting the environment, and advancing scientific knowledge.

Dear Readers,

Thank you for choosing "Bacteriology Uncovered: The Secrets of Microbial World Revealed." Your interest in exploring the fascinating world of bacteriology is truly appreciated.

If you have enjoyed reading this book, I kindly ask you to take a moment to share your thoughts by leaving a review. Your feedback is invaluable and helps other readers discover the insights and knowledge contained within these pages.

Here are a few points you might consider including in your review:

- What did you find most interesting or insightful about the book?

- How has the information provided changed or enhanced your understanding of bacteriology?

- Were there any specific chapters or sections that stood out to you?

- Any suggestions for improvement or additional topics you would like to see covered?

Your honest opinion will not only assist me in improving my work but also guide future readers in their decision-making process.

Thank you once again for your support and for being a part of the journey to uncover the secrets of the microbial world.

Best regards,

Bhupen Thapa

www.ingramcontent.com/pod-product-compliance
Lightning Source LLC
Chambersburg PA
CBHW072021230526
45479CB00008B/314